AF885625

WHY

DO WE

CALCULATE

BACKWARDS?

LEFT TO RIGHT IS BETTER

Kenneth R. Williams

Published by Inspiration Books, 2010,
Kensglen, Nr Carsphairn, Castle Douglas, DG7 3TE, Scotland, U.K.

ISBN: 978-1-902517-23-0

© K. R. Williams 2010

http://www.vedicmaths.org

CONTENTS

INTRODUCTION

Most people calculate starting with the right-hand figure – the units figure.

Why? Because that's what they were taught in school and that's what everyone else does.

But do you realise that you are calculating backwards when you calculate from right to left?

Think about it. You are reading this text from left to right.

Numbers too are read from left to right. You **read** and **pronounce** the number:

345

as 'three hundred and forty five'. That is to say you pronounce the number from left to right: first the hundreds, then the tens and then the units.

And you **write** the number from left to right.

For these reasons calculating from **left to right is actually more natural**, and once you get used to it, it's easier too. There is absolutely nothing stopping you from calculating left to right.

Also, **the most significant and important figure in a number is the one on the left**, because it has the biggest value. For example, in the number 345, the 3 is the most important figure as it represents the hundreds; the others are only tens and units.

If you have to add, or multiply, two 6-figure numbers and you want only the first two figures of the answer, you will have to do the whole calculation if you start at the right.

But if you start from the left you can do just the work that is needed to answer the question and no more: so it's much more efficient.

We add, subtract and multiply from right to left, but **we divide from left to right**. Apart from being rather confusing (especially for children) it means that we cannot combine calculations.

Starting at the left **we can combine calculations** (this is discussed more fully in the Conclusion to this book).

In fact all arithmetic operations can be carried out from left to right and this has huge advantages, as you will see.

People often tell me they calculate from left to right, but they were never taught it. They worked it out for themselves because they wanted, or needed, to be able to calculate efficiently and fast. In fact most rapid mental calculators work left to right.

Although hand-held calculators are in widespread use nowadays there are many situations when quick calculations are needed: checking your change in the shop, closing a deal at the lunch table or on the phone, or making a transaction in the stock exchange. These need to be done quickly and accurately and maybe approximately. In such situations left to right is essential.

And apart from wasting your time and energy doing unneeded work, calculating from **right to left may be impossible** to do anyway (see the Conclusion).

So for all the above reasons, left to right really is the preferred direction in which to calculate.

WHY CALCULATE RIGHT TO LEFT?

So why do we calculate from right to left? Why are children taught to work from the right?
First, working from the right you know you are starting with the units, which is nice and easy. Whereas if you start on the left you may be finding hundreds, thousands etc. depending on the numbers and what you are doing with them. And after the units you have the tens, hundreds and so on: a neat progression.

The second answer to the question is tradition. We have always calculated from the right: our parents did it, our children do it. It is the accepted way. Traditions change only very slowly.

And maybe it was not realised that it is possible to calculate left to right with such ease and with such advantages. No alternative to right to left has appeared until now.

USING THIS BOOK

So the aim of this book is to show how to calculate from left to right and to highlight the advantages.

You don't have to unlearn the right to left method you have been taught – you will naturally use it less and less when you start to use the left to right method.

But you do need to practice, because the old methods are ingrained in us and so they seem easy. For this reason plenty of exercises are included in this book. Gradually left to right will become easy, and easier than the opposite method. And then new areas and new possibilities will begin to open up.

This book is in two parts. First you will see how to calculate **mentally** from the left. As already mentioned, left to right is ideal for mental computations.

You will be taken through addition, multiplication and subtraction from left to right.

In the second part you will see how to do **written** calculations from left to right. This is similar to the mental methods shown in the first part, with a surprising but simple way of handling the carry figures. And it's also very similar to right to left working, but has the advantage that you do no unnecessary calculations. So it's far more efficient and quicker.

The two parts are fairly self-contained, so that you can skip the mental part if you are only interested in the written methods. But for this reason you may skip a few of the examples in Part 2 if you did Part 1.

This book is for everyone: whether you are a child or an adult, whether you struggle with arithmetic or find it easy. It proceeds in easy stages. You are advised not to use any methods you may already rely on: try to follow the text and use only the methods shown. If your maths is not too good, read this book slowly and carefully. If you rush it you are likely to give up early.

ABOUT THE AUTHOR

My own interest in calculating from left to right started in 1981 when I was asked to come up with a topic for a project during the final year of my Astronomy degree at University College London. I had been studying Vedic Mathematics since 1971: this is an amazing ancient system of mathematics from India in which answers are generally obtained in one line (see Chapter 3). I decided my project would be 'Applications of Vedic Mathematics in Astronomy'. My supervisor, Dr McNally, knew nothing of Vedic Mathematics and told me that if I didn't come up with something specific in a week he would give me a subject for my project himself.

INTRODUCTION

Earlier in the Astronomy course we had been introduced to Kepler's Equation, an important equation in calculating the positions of planets. This equation is however very difficult to solve and solving it without a calculator would never be attempted nowadays.
But I found a way of obtaining the first figure and using it to get the second figure, and then the third figure, and so on, without using a calculator. My supervisor was impressed and allowed me to continue. Not only could I solve Kepler's Equation digit by digit from left to right, but I was able to devise a method of predicting eclipses by arriving at an equation which I was able to similarly solve digit by digit from the left [2].

My project received very high marks and was to be published, and I was encouraged to continue my studies in this direction.

Later I went on to show how to solve other problems by this method. But I soon also realised that the techniques could be applied at elementary levels too, and I combined these with my continuing research into Vedic Mathematics. Together, the left to right method, with the one-line Vedic techniques, give a new approach to calculation, both mental and written.

As I used and taught the Vedic system, which I have been doing for about thirty years, I found the left to right methods were becoming more prominent. People would tell me they had found that calculating left to right was so much easier for them. Or they would tell me that they have already taught themselves one of the left to right techniques. Children too told me they liked and used the left to right method.

So although left to right calculation features in other books I have written it has never been the main focus. But as it is a subject which has so much to offer it seems right now to give an introduction to this powerful technique.

We do not go into advanced work here: you can see that in other books (References 1, 2, 3, 4). Here we simply show the basics. It is up to the reader to give these methods a fair hearing by trying them out. At first it may well seem harder than what you have learnt, but that's only because you are thoroughly accustomed and habituated to what you were taught in school. Once you can calculate left to right with some facility you will feel comfortable with it and want to continue using it.

PART 1 – IN YOUR HEAD

(Mental Calculations)

There are many advantages to mental maths. Of course there is the practical advantage of not being reliant on a calculator, being able to make quick and accurate decisions, check your change and so on.

But being mentally adept with numbers will give you more confidence too. And it will improve your memory and mental agility generally (see the Introduction to Reference 5).

1 ADDITION

Let's start with the simple addition of two 2-figure numbers.

To find, say, **34 + 42** you can easily add from left to right by adding first the tens to get 3 + 4 = 7 and then the units to get 4 + 2 = 6.

That gives the answer as 76.

34 + 42 = 76.

And it's very easy because there are no carries: the units add up to a single figure and so there is no carry from the units over to the tens.

Now suppose you have **34 + 58**.

You see that the tens total is 3 + 5 = 8.
Remembering this **8**, add the units: 4 + 8 = 12.

So the 1 in this 12 is a carry figure, and the 8 we got for the tens must be increased by this 1 to give **9** as the number of tens in the answer.

And because we had 12 for the units, the 2 here goes onto the end of the 9 to give the answer as 92.

34 + 58 = 92.

In other words you add the tens, remember it, and add the units. Then increase the tens by 1 if there is a carry coming over from the units, and attach the last figure of the units on the end.

We can show it like this: **$34 + 58 = 8_12 = 92$**.

For the purposes of this book we will use the notation shown on the line above, to show the mental method. So a carry figure is written as a subscript.

This may seem very simple (I hope so!), but this is the key to the whole thing so be sure to take it in, and practice it.

" The advancement and perfection of Mathematics are intimately connected with the prosperity of the state."

Napoleon

Another example: **76 + 58**.

Think: 7 + 5 = 12 and 6 + 8 = 14.
So the 1 in the 14 is carried over to the 12 to make 13, giving the answer of 134.

76 + 58 = $12_{1}4$ = **134**.

Now practice a few of these, to begin forming the habit of working left to right.

Practice

a 58 + 38 **b** 77 + 45 **c** 47 + 49

d 94 + 49 **e** 88 + 88 **f** 67 + 44

Answers: **a** 96 **b** 122 **c** 96 **d** 143 **e** 176 **f** 111

LONGER NUMBERS

Longer numbers can be tackled in just the same way.

Suppose you need **187 + 444**.

Add the hundreds first, then the tens, and finally the units.

The hundreds gives 1 + 4 = 5 so **remember 5**.
Now the tens: 8 + 4 = 12.
That means we carry the 1 in the 12 over to the 5 in the hundreds place and append the 2.

So now we have 62.

Remember 62, and add the units: 7 + 4 = 11.

The first 1 in this 11 is carried over to the 62 to make it 63 and the 1 is appended, giving 631.

So **187 + 444 = 631**.

We build up the answer step by step from the left, carrying 1 whenever we get a 2-figure total.

The first two steps in the last example combine like this: $5_12 = 62$.
And this is combined with the last step like this: $62_11 = 631$.

It can all be done very quickly and very efficiently with a little practice.

Take **777 + 753** next.

The first combination is $14_{1}2$ = 152 (remember 152).
Then combine this with the units: $152_{1}0$ = 1530.

So **777 + 753 = 1530**.

Practice some 3-figure additions now. Be sure to do the whole thing in your head.

Practice

a 567 + 358 **b** 279 + 345 **c** 876 + 555

d 543 + 282 **e** 753 + 419 **f** 777 + 777

Answers: **a** 925 **b** 624 **c** 1431 **d** 825 **e** 1172 **f** 1554

At the age of three Gauss corrected his father when he was calculating the wages due to one of the labourers in his charge. On checking the account the figure named by young Gauss was found to be correct.
When he was ten Gauss' teacher asked his class to add up all the numbers from 1 to 100. Gauss immediately wrote down the answer on his slate and later when all the slates were checked the teacher was astonished to find that only Gauss had given the correct answer.

Longer additions can be done in just the same way. Simply start at the left and add in each position, taking up the carry where it comes up.

So for **8,585 + 6,671** you get **15,256**.

The steps are: $14_11 = 151$,
$151_15 = 1525$,
15256

This is the method that many people teach themselves, because it's more natural and more efficient. And it seems pointless to start by looking at the units, which are the least important figures.

This is especially true if you only want an estimate of the answer: an approximation.

"Mathematics is the Queen of the Sciences and Arithmetic the Queen of Mathematics"

Carl F. Gauss *

* Carl Friedrich Gauss (1777-1855) is usually reckoned as one of the three greatest mathematicians ever, along with Archimedes and Newton.

APPROXIMATIONS

Note that if you don't need the exact answer you can just stop whenever you like.

When finding **8,585 + 6,671**, if you only want the thousands total you just add the thousands, which is 8+6 = 14.

But then a glance at the hundreds shows that this will be increased by 1 (because 5+6 = 11, so there will be a carry).

So to the nearest thousand **8,585 + 6,671 = 15,000**.

Find **57,555 + 8,844 + 67** to the nearest thousand.

Here we can ignore the 67 at the end as it will not affect the thousands.

Adding the thousands we get 57 + 8 = 65.

And there will be another 1 from the hundreds as 5 + 8 = 13.

So adding that 1 onto 65 we get:
57,555 + 8,844 + 67 = 66,000 to the nearest thousand.

Similarly, for **643 + 743** to the nearest hundred we initially get 6+7 = 13, so that's 1300 or 1,300.

Now look at the tens figures.

We have 4+4 = 8.

Although there is no carry here we should still increase the 13 hundreds to 14 hundreds.

This is because we should 'round up': 1,380 is closer to 1,400 than to 1,300.

So **643 + 743 is 1,400** to the nearest hundred.

In fact the well-known rule is that if the tens total is 5, 6, 7, 8 or 9 you increase the hundreds by 1.

Similarly if you want a number to the nearest thousand you increase that total by 1 if the hundreds total is 5, 6, 7, 8 or 9. And so on.

Tommy arrives at school with a cup – the sort you win at a tournament. The teacher says "what's that?" Tommy replies "I won it in a maths competition". Surprised the teacher asks "how come?" Tommy replies "they asked me what is two sevens. I said 13 and got third prize!"

Practice

Find to the nearest hundred:

a 568 + 857 **b** 717 + 625 **c** 947 + 547

Find to the nearest thousand:

d 5,348 + 3,847 **e** 17,444 + 18,333 **f** 2,345 + 9,412

Answers: **a** 1400 **b** 1300 **c** 1500 **d** 9,000 **e** 36,000 **f** 12,000

" Like the crest of the peacock,
like the gem on the head of a snake,
so is mathematics at the head of all knowledge."

*from the **Jyotish Vedanga***

2 SHORT MULTIPLICATION

Now let's move on to multiplication. This is done in just the same way as addition.

If you want **23 × 3** you can multiply each digit in 23 by 3 and you can do it from left to right.

So **23 × 3 = 69**.

But **23 × 4** involves a carry.

First you get 2×4 = 8 and remember **8**.

Then 3 × 4 = **12** and the 1 here is carried over to the 8 from before to give **92** as the answer.

We can show this: **23 × 4 =** $\mathbf{8_12}$ **= 92**.

So **23 × 4 = 92**.

This is just like the addition, except we're now multiplying.

Similarly **66 × 9** = 54_54 = **594**.

It's just a matter of getting used to this simple technique of multiplying from the left and adjusting for any carries that come over from the next step.

The answer is built up in stages from the left.

And **77 × 7** = 49_49 = **539**.
Here you mentally increase 49 by 4 to get the 53.

Practice

a 42 × 3 **b** 57 × 3 **c** 77 × 5

d 94 × 7 **e** 88 × 8 **f** 67 × 3

Answers: **a** 126 **b** 171 **c** 385 **d** 658 **e** 704 **f** 201

LONGER NUMBERS . . .

For longer numbers the same method applies.

Take **357 × 8**.

First 3×8 = 24, remember **24**.

Then 5×8 = 40, so combining this 40 with the 24 we get 24_40 = 280.

Remember **280**.

Finally 7x8 = 56 which we combine with the 280: 280_56 = **2856**.

So **357 x 8 = 2856**.

Take another example: **487 × 6**.

First 4×6 = 24, remember **24**.

Then 8×6 = **48**. So combining: 24_48 = 288. Remember **288**.

Then 7×6 = **42**. And combining: 288_42 = **2922** (your 288 is mentally increased by the carried 4 to make 292).

So **487 × 6 = 2922**.

Practice

a 532 × 4 **b** 777 × 3 **c** 567 × 7

d 954 × 8 **e** 484 × 6 **f** 123 × 9

Answers: **a** 2128 **b** 2332 **c** 3969 **d** 7632 **e** 2904 **f** 1107

. . . AND LONGER NUMBERS

Like addition, we can handle numbers of any size like this.

So for **5913 x 9** we first get 45, then 81.

Combining these: 45_81 = 531. Remember **531**.

Next we get 9 to combine with 531, which is just **5319** (there's no carry here).

Finally we get 27 to combine with 5319: 5319_27 = **53217**.

So **5913 x 9 = 53217**.

Practice

a 4321 × 8 **b** 3366 × 7 **c** 9999 × 9

Answers: **a** 34568 **b** 23562 **c** 89991

APPROXIMATIONS

And if we don't need the exact answer we can stop whenever we have the accuracy needed.

So to find **3,546 x 6** to the nearest thousand we see that 3x6 = 18, so that's 18,000.

To be more accurate though we can see if there is a carry to add to this in the next step.

So multiplying the second figure by 6 we get 5x6 = 30, and this tells us that a 3 will need to be carried to the 18 from the first step, giving 21 thousands.

So **3,546 x 6 = 21,000** to the nearest thousand.

Similarly **7,654 x 3** gives 21,000 initially, looking at only the first figure.

Looking at the second figure: 6x3 = 18, so $21{}_{1}8$ = 228. And this is 22,800 so we must round this up to **23,000**.

Practice

Find to the nearest hundred:
a 582 × 3 **b** 717 × 6 **c** 947 × 5

Find to the nearest thousand:
d 5,348 × 7 **e** 17,444 × 3 **f** 2,825 × 9

Answers: **a** 1700 **b** 4300 **c** 4700 **d** 37,000 **e** 52,000 **f** 25,000

Sometimes though we need to go a step further if we want to be precise in the answer to the nearest hundred, thousand etc.

If you need **3,456 x 6** to the nearest thousand, you first get 18.

The next step is 24 and so gives 18_24 = 204.

So you may think that since that 4 here is small the answer will be 20,000.

However, there may be a carry in the third step that will increase that 4 to a larger digit.

The third step is 5x6 = 30 so that 3 will be added to that 4, making the first three figures 207.

And this means that we really get **21,000**.

Practice

Find to the nearest thousand:

a 5,681 × 3 **b** 2,345 × 9 **c** 3,579 × 5

Answers: **a** 17,000 **b** 21,000 **c** 18,000

3 VEDIC MATHEMATICS

Before we go on to subtraction and the so-called 'long' multiplication we will look at the ancient system of mathematics known as "Vedic Mathematics".

The reconstruction of this system of mathematics from the Indian Vedic texts is the work of Sri Bharati Krsna Tirthaji (1884-1960). He points out that all of mathematics, pure and applied, is based on sixteen basic formulae or Sutras and he initially wrote sixteen books, one on each Sutra, showing applications of those Sutras.

The Sutras are given in word form and have many applications. They express in mathematical form certain basic ways in which we naturally use our mind (see the Appendix in Reference 8,). One formula says: *All from 9 and the Last from 10* and another is *Vertically and Crosswise*.

> *"exceedingly tough mathematical problems (which the mathematically most advanced present day Western scientific world had spent huge lots of time, energy and money on and which even now it solves with the utmost difficulty and after vast labour involving large numbers of difficult, tedious and cumbersome 'steps' of working) can be easily and readily solved with the help of these ultra-easy Vedic Sutras (or mathematical aphorisms).... in a few simple steps and by methods which can conscientiously be described as mere 'mental arithmetic'."*
>
> ***Bharati Krsna Tirthaji***

The sixteen books just mentioned were lost however and Bharati Krsna wrote an introductory volume in 1958 which was published in 1965 [6]. Since then others including myself have sought to expand and extend the contents of this remarkable book.

But Vedic Mathematics has been misunderstood in recent years: while some have gone on to apply the Vedic methods in various areas of mathematics and computing with considerable success, others, who have not really looked closely at what Vedic Mathematics has to offer, have denounced the system.

Time will no doubt show that the beautifully simple and unified Vedic system is far better than what is taught conventionally at present.

In the Vedic system we can multiply and divide any numbers of any size and get the answer in one line: such is the simplicity and power of the Vedic Maths.

In this book we will introduce Vedic subtraction and multiplication. If you are interested in further Vedic methods please consult one of the books devoted to this system (e.g. Refs 5, 6, 7, 8).

The Vedic system has both general and special methods, which means that in addition to the general methods, special methods are also catered for. So, for example, there is the general multiplication method that gives the product of any two numbers in one line. But there are especially easy methods for special cases, like products where the numbers are close to 100, 1000 etc.

These are not covered here, except for the grouping method, so please see another text for a fuller exposition of the special methods.

The combination of special and general methods is a great advantage in the teaching and application of mathematics. It brings flexibility and choice so that instead of having just one method which must be applied, you can choose and look out for the characteristics of those special cases. So maths becomes an interesting and inventive activity instead of being rigid and therefore tedious.

In the opinion of this author Vedic Mathematics is the solution to the continuing problem facing the teaching of mathematics: its poor reputation amongst children and adults alike, the overuse of hand-held calculators and the low abilities in calculation generally.

Mathematics is fun and totally consistent and reliable and we are certainly not fully developing the abilities of which our children are capable.

> *"Even the highest and farthest reaches of modern Western mathematics have not yet brought the Western world even to the threshold of Ancient Indian Vedic Mathematics."*
>
> **Professor Augustus De Morgan**

4 SUBTRACTION

There are two wonderful Vedic subtraction methods, both of which work from left to right. Normally subtraction is carried out from right to left and many children find it difficult to master. But both Vedic methods are really easy, as you will see.

METHOD 1 – "ALL FROM 9 AND THE LAST FROM 10"

The Vedic formula *All from 9 and the Last from 10* subtracts numbers from the next highest base.

By 'base' we mean a power of 10: so 10, 100, 1000 etc. are base numbers.

If you apply *All from 9 and the Last from 10* to, say, **357** it means you take each digit in 357 from 9, but you take the last one from 10.

1000 –	**3**	**5**	**7**
	from 9	from 9	from 10
=	**6**	**4**	**3**

So 3 from 9 is 6,
5 from 9 is 4,
and 7 from 10 is 3.

And so **1000 – 357 = 643**.

Similarly, **10,000 – 8,742 = 1,258**.

You take the 8, 7 and 4 from 9 and you take the 2 from 10.

Practice

a 1000 – 281 **b** 1000 – 777 **c** 10,000 – 3,456

d 100 – 38 **e** 1000 – 909 **f** 100,000 – 67,654

Answers: **a** 719 **b** 223 **c** 6,544 **d** 62 **e** 91 **f** 32,346

This kind of subtraction, where you're subtracting from a base number, is normally the hardest kind of subtraction, because it involves repeated borrowing. But the Vedic method shows that it's really the easiest.

That the method is valid can be seen by finding 1000 – 357 by the usual method:

$$\begin{array}{r} 1\,0\,0\,0 \\ -\ \underline{3\,5\,7} \\ \underline{} \end{array} \quad \text{becomes} \quad \begin{array}{r} {\scriptstyle 9\ \ 9} \\ \not{1}^1\,\not{0}^1\not{0}^1 0 \\ -\ \underline{3\,5\,7} \\ \underline{} \end{array}$$

In which we keep borrowing and reducing by 1, and the final result is that we take each digit in 357 from 9 and the last from 10!

NUMBERS ENDING IN ZERO

If you apply the formula to **1000 – 670** you get $32_{1}0$ which equals **330**.
There is however a better way for subtracting numbers that end in a zero or zeros. And that is to apply the formula to the number without the zero and put the zero on afterwards.

So for **1000 – 670** you apply *All from 9 and the Last from 10* to 67 to get 33, and then put the zero on, to get **330**.

Similarly with **10,000 – 4,400** you apply the formula to 44 to get 56 and put two zeros after it (because 4400 ends with two zeros).
That gives **10,000 – 4,400 = 5600**.

Practice

a 1000 – 280 **b** 10,000 – 6690 **c** 10,000 – 3,400

Answers: **a** 720 **b** 3,310 **c** 6,600

MONEY

This method has useful practical applications because if you buy an item priced at **$6.46** in the shop and you pay with a **$10** note, you can simply apply this formula to find the expected change.

$10 – $7.46 = $2.54 (take 7 and 4 from 9, and take 6 from 10).

And **$100 – $53.68 = $46.32** (take 5, 3 and 6 from 9, and 8 from 10).

Practice

a $10 – $3.21 **b** $10 – $8.65 **c** $100 – $37.32

Answers: **a** $6.79 **b** $1.35 **c** $62.68

"Number rules the universe"

Pythagoras

ADDING ZEROS

In all these examples the number of digits in the subtracted number is the same as the number of zeros in the base number you are subtracting from.

For example, in 1000 – 777 there are 3 zeros, and there are 3 figures in 777.

Suppose you need: **10,000 – 777**.

Here there are 4 zeros in 10,000 and there are 3 figures in 777.

To apply the same method we can make 777 into a 4-figure number by simply putting a zero in front like this:

10,000 – 0777.

Then apply the formula to 0777.

That gives: **10,000 – 0777 = 9223**.

Of course this is done mentally (and you may have to mentally add more than one zero, as in the next example).

Now look at: **100,000 – 246**.

Here 100,000 has 5 zeros and 246 is a 3-figure number, so we make 246 into a 5-figure number, by putting two zeros on the left of it.

So you think: **100,000 – 00246 = 99,754** (taking 0, 0, 2, 4 from 9, and 6 from 10).

Practice

a 1000 – 77 **b** 10,000 – 77 **c** 10,000 – 777

d 1000 – 7 **e** 10,000 – 990 **f** 100,000 – 54

Answers: **a** 923 **b** 9923 **c** 9223 **d** 993 **e** 9010 **f** 99946

MULTIPLE OF A BASE NUMBER

Another variation is where you are subtracting from a **multiple** of a base number, like 200 or 3000 or 70,000 and so on (rather than 100, 1000 or 10,000).

Suppose you want **7,000 – 161**.

Because the 161 will come off one of those seven thousands there will be 6 thousands left.

So **7,000 – 161 = 6,839**.

That is, the 7 is reduced by 1 to give 6, and the *All from 9 and the Last from 10* formula is applied to 161 to give 839.

Similarly, with **300 – 57** you get **243**.

The 3 is reduced by 1 to 2, and the formula is applied to 57 to give 43.

You may have to put in a zero as well as do this reducing of the first figure....

For **40,000 – 852** note that 40,000 has 4 zeros and that 852 is a 3-figure number.

That means putting a zero before the 852, so:

40,000 – 852 = 40,000 – 0852 = 39148.

The 4 is reduced by 1 to give 3 as the first figure of the answer, and the formula is applied to 0852 to give 9148.

Practice

a 4000 – 333 **b** 500 – 88 **c** 60,000 – 5,432

d 7000 – 38 **e** 90,000 – 157 **f** 40,000 – 57

Answers: **a** 3667 **b** 412 **c** 54568 **d** 6962 **e** 89843 **f** 39943

Although this method does lead to a completely general subtraction method, we will leave it for now, until Chapter 10.

"the construction of the physical and moral world alike is based on eternal numbers"

St. Augustine

METHOD 2

Now let's look at the other left to right mental subtraction method.

Suppose you want **82 – 37**.

Looking at the tens digits we see 8 – 3 = 5.

But a glance at the units shows that 2 – 7 can't be done. So the first figure is not 5 but one less: **4**.

Now that 1 which we took off the 5 is given to the 2 in 82 to make it into 12, so you are now thinking like this:

$\mathbf{8_{1}2 - 37 = 4?}$

And finally in the units column, 12 – 7 = 5 and the answer is 45.

So **82 – 37 = 45**.

In this method you start at the left, subtract, and give this as the first figure if the subtraction in the next column **can** be done.

If it **cannot** be done you reduce the first answer figure by 1, carry 1 to the right, and then subtract in the second column.

Similarly, with **74 – 46** you look at the tens figures and subtract.

You get 7 – 4 = 3.

But since the subtraction in the units column cannot be done we give 2 as the first figure of the answer (1 less than 3), and append that 1 to the 4 in 74:

7_14 – 46 = 2?

Then, in the units we have 14 – 6 = 8, so the answer is 28.

74 – 46 = 28.

Practice

a 64 – 38 **b** 44 – 29 **c** 82 – 47

d 91 – 38 **e** 55 – 27 **f** 73 – 57

Answers: **a** 26 **b** 15 **c** 35 **d** 53 **e** 28 **f** 16

And this can be extended to numbers of any length: Just subtract in each position starting at the left, but always check the next column and carry a unit into it if the subtraction there cannot be done.

Let's subtract a pair of 3-figure numbers next.

In **753 – 482** we get 7 – 4 = 3.

But since the middle column cannot be done (5 – 8) we give 2 as the first figure and carry 1 to the 5 to make it 15.

Think of it like this: **$7_1$53 – 482 = 2??**

Now subtracting in the middle column: 15 – 8 = 7. And we don't have to reduce this because the last column can be done: 3 – 2 = 1.

So we have: **753 – 482 = 271**.

After a little practice it is very easy and fast.

Practice

a 634 – 378 **b** 704 – 323 **c** 582 – 347

d 333 – 177 **e** 5665 – 2777 **f** 7003 – 5775

Answers: **a** 256 **b** 381 **c** 235 **d** 156 **e** 2888 **f** 1228

THE SAME FIGURES IN A COLUMN

There is one more consideration, because it may happen that you glance forward to decide whether to reduce or not and find that in that column the figures are in fact the same.

In such a case we look at the **next** column.

For example: **934 – 438**.

In the left column 9 – 4 = 5.

Do we need to reduce this 5?

Looking in the middle column we see that both figures are 3.

So look at the last column. Here we see we cannot do 4 – 8.

That means we do need to reduce that 5 to a 4.

$\mathbf{9_{1}34 - 438 = 4??}$

The next step is 13 – 3, which is 10. But this is reduced to 9, and 1 is given in the last column:

$\mathbf{9_{1}3_{1}4 - 438 = 49?}$

Finally subtracting in the last column: 14 – 8 = 6.

So **934 – 438 = 496**.

Practice

a 564 – 368 **b** 444 – 249 **c** 8282 – 1887

Answers: **a** 196 **b** 195 **c** 6395

5 VEDIC MULTIPLICATION

The Vedic multiplication method is amazingly simple and effective.

It can give the product of any two numbers:

- using a simple pattern,
- which is easy to explain,
- in one line,
- from right to left or from left to right.

It is also reversible so that any two numbers can be divided in one line [7].

The same technique can be used for multiplying algebraic expressions, and this can also be reversed to carry out algebraic divisions [8].

Nor is this all. This multiplication method simplifies for the squaring of numbers, and this too can easily be reversed for finding one-line square roots, and extended to finding solutions to quadratic and other equations [1,4].

Such is the range and versatility of the Vedic system: simple but powerful.

We will not cover all these topics here though. Only products of numbers using the simple pattern that gives the answer in one line from left to right. Please see the other books for the other topics.

The Vedic formula that applies here is: *Vertically and Crosswise*.

VEDIC MULTIPLICATION

This follows on neatly from multiplication of a number by a single figure, covered earlier in this book. So here is a brief reminder of the method we used then.

> In finding **67 x 8** we multiply the 6 by 8 to get 48.
>
> Then we multiply the 7 by 8 to get 56.
>
> And we combine these mentally: $\mathbf{48_{5}6 = 536}$.

That is to say the tens figure of the second product (the 5) is mentally added to the answer to the first product.

"That vast book which stands forever open
before our eyes, the universe,
cannot be read until we have learned the
language and become familiar with the
characters in which it is written.
It is written in mathematical language, without
which means it is humanly impossible to
comprehend a single word.

Galileo Galilei

THE VERTICAL / CROSSWISE / VERTICAL PATTERN

Let's start with a simple example.

Suppose we want **21 x 23**.

$$\begin{array}{r} 2\ \ 1 \\ \underline{2\ \ 3} \times \\ \underline{4\,8\,3} \end{array}$$

There are three steps, A, B and C:

A. multiply vertically in the left-hand column: 2 × 2 = 4, so 4 is the first figure of the answer.

$$\begin{array}{r} 2\ \ 1 \\ \times\ \underline{2\ \ 3} \\ \underline{4\ \ \ \ \ } \end{array}$$

B. multiply crosswise and add: 2 × 3 = 6, 1 × 2 = 2, 6 + 2 = 8, so 8 is the middle figure of the answer.

$$\begin{array}{r} 2\ \ 1 \\ \times\ \underline{2\ \ 3} \\ \underline{4\ 8\ \ } \end{array}$$

C. multiply vertically in the right-hand column: 1 × 3 = 3, 3 is the last figure of the answer.

$$\begin{array}{r} 2\ \ 1 \\ \times\ \underline{2\ \ 3} \\ \underline{\mathbf{4\ 8\ 3}} \end{array}$$

This simple vertical / crosswise / vertical pattern is all we need.

Take another example: **14 x 21**.

$$\begin{array}{r} 1\quad 4 \\ \underline{2\quad 1} \times \\ \underline{2\ 9\ 4} \end{array}$$

A. vertically on the left: 1×2 = **2**,
B. crosswise: 1×1 = 1, 4×2 = 8 and 1+8 = **9**,
C. vertically on the right: 4×1 = **4**.

This is of course very easy and straightforward, and we should now practice this vertical and crosswise pattern to establish the method.

Practice

a 22 × 31	**b** 21 × 31	**c** 21 × 22	**d** 22 × 13
e 61 × 31	**f** 32 × 21	**g** 31 × 31	**h** 13 × 13

Answers: **a** 682 **b** 651 **c** 462 **d** 286 **e** 1891 **f** 672 **g** 961 **h** 169

" A mathematician, like a painter or poet, is a maker of patterns."

G H Hardy

WITH CARRIES

The previous examples involved no carry figures, so let us consider this next.

Suppose we need **23 x 41**.

$$\begin{array}{cc}2 & 3\\ \underline{4} & \underline{1} \ \times\\ \multicolumn{2}{c}{\underline{943}}\end{array}$$

A. First we multiply vertically on the left: 2 x 4 = 8. **Remember 8**.

B. Next multiply crosswise and add:
2×1 + 3×4 = 14.

Now, as previously when we were adding and multiplying by a single figure from left to right, we combine the 8 and the 14 by carrying the 1 in the 14 to the 8 from the first step.

So 8_14 = 94. **Remember 94**.

C. Finally we multiply vertically on the right:
3 x 1 = 3.
And appending this to 93 we get **943** as the answer.

So **23 x 41 = 943.**

So it's the same as before, but with the occasional carry.

Another example: **23 x 34**.

$$\begin{array}{r} 2\ \ 3 \\ \underline{3\ \ 4} \times \\ \underline{782} \end{array}$$

A. We get 2 x 3 = 6. **Remember 6**.
B. Then 2x4 + 3x3 = 17.
   Combining with the 6: $6_{1}7$ = 77. **Remember 77**.
C. 3 x 4 = 12.
   Combining with the 77: $77_{1}2$ = **782**.

You can now multiply any two 2-figure numbers together in one line. With practice you will find that you can multiply the numbers mentally and fast.

Practice

a 21 × 47	**b** 23 × 43	**c** 23 × 24	**d** 44 × 53
e 45 × 33	**f** 74 × 38	**g** 38 × 38	**h** 38 × 88

Answers: **a** 987 **b** 989 **c** 552 **d** 2332 **e** 1485 **f** 2812 **g** 1444 **h** 3344

This simple multiplication pattern is easy to explain because the vertical product on the left is multiplying tens by tens and so gives the number of hundreds in the answer.

Similarly the crosswise step is multiplying tens by units and units by tens and so gives the number of tens in the answer, and the vertical product on the right is multiplying units by units and so gives the number of units.

Ten-year old Truman Henry Safford was asked:
"multiply in your head 365,365,365,365,365,365
by 365,365,365,365,365,365.
He flew around the room like a top, pulled his pantaloons over the top of his boots, bit his hand, rolled his eyes in their sockets, sometimes smiling and talking, and then seeming to be in an agony, until, in no more than one minute, said he,
133,491,850,208,566,925,016,658,299,941,583,225!"

In Appendix 1 you can see how a grouping technique allows use of this pattern for multiplying 3 and 4-figure numbers. And in Appendix 2 you can see how the pattern can be used to multiply 3, 4 and 5-figure numbers by a 2-figure number (the Moving Multiplier Method).

Although the vertical and crosswise pattern shown here can give all products, the pattern itself can be developed further for multiplying 3, 4 etc. figure numbers. See Appendix 3 for this.

PART 2 – ON PAPER

(Written Calculations)

The paper methods of left to right calculation are basically the same as for mental calculations, here we simply show the best way to set it out, and the rather surprising way of dealing with the carry figures.

If you have read Part 1 carefully you may not need to go through all the examples that follow. This especially applies to Chapter 7.

Written calculations are really essential for more advanced work and this section of the book sets things up for the more advanced applications [1,4].

6 ADDITION

Now, rather than calculating **mentally** let's look at how to **write** calculations from left to right.

And let's start with addition.

But first we will just remind ourselves about adding from right to left because something needs to be pointed out about that which is relevant to the left to right process.

To find **56 + 78** we set it out like this:

$$\begin{array}{r} 5\ 6 \\ +\ \underline{7\ 8} \\ \underline{\quad\ } \end{array}$$

And we add the figures on the right: 6 + 8 = 14.
Then we put down 4 and carry 1, like this:

$$\begin{array}{r} 5\ \mathbf{6} \\ +\ \underline{7\ \mathbf{8}} \\ \underline{4} \\ {\scriptstyle 1\ \ } \end{array}$$

At the next step we find 5 + 7 = 12 and we add the carried 1 to get 13, which we put down. Now we have:

$$\begin{array}{r} \mathbf{5}\ 6 \\ +\ \underline{\mathbf{7}\ 8} \\ \underline{13\ 4} \\ {\scriptstyle 1\ \ } \end{array}$$

Now note this: that 1 which we carried was really 10, because it was in the 14, and yet it was counted as 1 in the next column. Its value became ten times **smaller**.

This may seem a minor point, but when we work in the opposite direction next, the reverse will happen: the carry becomes ten times **bigger**.

Suppose you want **56 + 78** again.

Set the numbers out in the usual way:

```
  5 6
+ 7 8
  ___
```

Adding the left column first: 5 + 7 = 12, put down 1 and carry 2, like this:

```
    5 6
+   7 8
  -----
  1
    2
```

Now add the right-hand column: 6 + 8 = 14, and add on the carried 2 **as 20**.

(Because every carry number when working left to right becomes ten times bigger.)

So 14 + 20 = 34, which we put down.

```
    5 6
+   7 8
  -----
  1 3 4      The same answer as before.
    2
```

The important point is: **Every carry figure when working left to right becomes ten times bigger**.

Actually we already do this in division, which is always carried out from left to right.

If, for example, we **divide 54 by 3**:

$$\begin{array}{l} 1 \\ 3\overline{)5\;{}_{2}4} \end{array}$$

The first step is to divide 5 by 3, which is 1 remainder 2.
You can see the remainder 2 shown as a subscript, and we then divide 24 by 3 to get 8. So 54 ÷ 3 = 18.

But notice that the subscript **2 is counted as 20** to make that 24.

So for **85 + 86**:

$$\begin{array}{r} 8\;5 \\ +\quad 8\;6 \\ \hline 1\;7\;1 \\ \hline {}_{6} \end{array}$$

Here we get 16 when adding the left column.
So put down 1 carry 6.
Add the right column to get 11.
Then add 11 + 60 (the carried 6 is counted as 60) to get 71 which you put down.

In the case of adding **38 + 49** we find the first total is a single figure, not a 2-figure number as before.

$$\begin{array}{r} \mathbf{3}\,8 \\ +\ \mathbf{4}\,9 \\ \hline 0 \\ \hline {\scriptstyle 7} \end{array} \qquad\qquad \begin{array}{r} 3\,\mathbf{8} \\ +\ 4\,\mathbf{9} \\ \hline 0\,8\,7 \\ {\scriptstyle 7} \end{array}$$

The 7 we get in adding the first column can be thought of as 07, so that we put down 0 and carry 7.

Then adding on the right we get 17, and adding the 7 (as 70) gives 87 as the answer.

Alternatively we can just leave that 0 out:

$$\begin{array}{r} 3\,8 \\ +\ 4\,9 \\ \hline 8\,7 \\ \hline {\scriptstyle 7} \end{array}$$

So **38 + 49 = 87**.

Practice

a $\begin{array}{r} 58 \\ +\ 38 \\ \hline \end{array}$ **b** $\begin{array}{r} 77 \\ +\ 45 \\ \hline \end{array}$ **c** $\begin{array}{r} 47 \\ +\ 49 \\ \hline \end{array}$

d $\begin{array}{r} 94 \\ +\ 88 \\ \hline \end{array}$ **e** $\begin{array}{r} 88 \\ +\ 44 \\ \hline \end{array}$ **f** $\begin{array}{r} 67 \\ +\ 49 \\ \hline \end{array}$

Answers: **a** 96 **b** 122 **c** 96 **d** 182 **e** 132 **f** 116

ADDING 3-FIGURE NUMBERS

Suppose you want **654 + 777**.

```
   6 5 4
+  7 7 7
  ______
```

Adding the left column first we get 13:

```
    6 5 4
+   7 7 7
  1
   3
```

In the middle column we get 12, and adding the carried 3 (as 30) we get 42.

So put down 4 and carry 2:

```
    6 5 4
+   7 7 7
  1 4
   3 2
```

Finally adding in the right column we get 11 to which we add the carried 2 (as 20) to get 31. Put down 31:

```
    6 5 4
+   7 7 7
  1 4 3 1
   3 2
```

Actually this is very simple and easy, but it's just the reverse of adding right to left that we are accustomed to.

It may seem strange, but with a little practice it will become quite natural.

It is also worth pointing out that this way we always put two figures down at the last step.

Practice

a $\begin{array}{r} 567 \\ +\ \underline{358} \end{array}$	**b** $\begin{array}{r} 279 \\ +\ \underline{345} \end{array}$	**c** $\begin{array}{r} 876 \\ +\ \underline{555} \end{array}$
d $\begin{array}{r} 543 \\ +\ \underline{282} \end{array}$	**e** $\begin{array}{r} 753 \\ +\ \underline{419} \end{array}$	**f** $\begin{array}{r} 777 \\ +\ \underline{777} \end{array}$

Answers: **a** 925 **b** 624 **c** 1431 **d** 825 **e** 1172 **f** 1554

The teacher asks "what is 7 times 6?"
Tommy replies "42"
The teacher says "very good! - And what is 6 times 7?"
Tommy replies "It's 24"

7 MULTIPLICATION

The same technique we just used for writing left to right additions can also be used for multiplication.

So to find **64 x 7**:

```
    6 4
x    7
  4
   2
```

We first find 6 x 7 = 42 and put down 4 carry 2 as shown above.

Now find 4 x 7 = 28.

So add the carried 2 (as 20): 28 + 20 = 48, which we put down:

```
    6 4
x    7
  4 4 8
   2
```

We simply use the same method as addition: remembering that any carry is ten times its value at the next step.

Notice that the multiplier (7 in the example above) is aligned at the left here, instead of at the usual position on the right.

Usually children are taught to line the units etc. up, under each other, and this is especially important for addition and subtraction. It makes more sense here to line up on the left, but of course if you prefer the usual way please do that.

Similarly for **55 x 5**:

$$\begin{array}{r l} & 5\ 5 \\ \times & \underline{5\ \ \ } \\ & \underline{\mathbf{2\ 7\ 5}} \\ & {\scriptstyle 5}\ \ \ \ \ \end{array}$$

We first get 25, put 2 carry 5.
Then we get another 25, and 25 + 50 = 75, giving 275 as the answer.

You can also write this as follows:

$55 \times 5 = 2_575 = 275$.

That is to say you can write it all on one line if you wish.

Practice

a $\begin{array}{r l} & 4\ 2 \\ \times & \underline{3\ \ } \end{array}$ **b** $\begin{array}{r l} & 5\ 7 \\ \times & \underline{3\ \ } \end{array}$ **c** $\begin{array}{r l} & 7\ 7 \\ \times & \underline{5\ \ } \end{array}$

d $\begin{array}{r l} & 9\ 4 \\ \times & \underline{7\ \ } \end{array}$ **e** $\begin{array}{r l} & 4\ 4 \\ \times & \underline{8\ \ } \end{array}$ **f** $\begin{array}{r l} & 7\ 3 \\ \times & \underline{6\ \ } \end{array}$

Answers: **a** 126 **b** 171 **c** 385 **d** 658 **e 352** **f 438**

LONGER NUMBERS

For longer numbers we do just the same thing: for example **345 x 8**.

```
   3 4 5
x    8
 2
   4
```

First we get 3 x 8 = 24.

Then 4 x 8 = 32 and 32 + 40 = 72, so put down 7 and carry 2:

```
   3 4 5
x    8
 2 7
   4 2
```

Finally 5 x 8 = 40 and 40 + 20 = 60, so put down 60:

```
   3 4 5
x    8
 2 7 6 0
   4 2
```

Similarly for **749 x 3**:

```
      7 4 9
x     3
   2 2 4 7
    1 2
```

And here's a 4-figure number....

Find **3456 x 8**:

```
     3 4 5 6
x    8
   2 7 6 4 8
    4 2  0
```

3 x 8 = 24, put 2 carry 4.
4 x 8 = 32, 32 + 40 = 72, put 7 carry 2.
5 x 8 = 40, 40 + 20 = 60, put 6 carry 0.
6 x 8 = 48, 48 + 0 = 48, put down 48.

Practice

a 632 × 4

b 777 × 3

c 567 × 7

d 954 × 8

e 464 × 3

f 4321 × 8

g 3366 × 7

Answers: **a** 2528 **b** 2331 **c** 3969 **d** 7632 **e** 1392 **f** 34568 **g** 23562

3-FIGURE TOTAL

Now look at **49 x 7**.

$$\begin{array}{rl} & 4\ 9 \\ \times & \ \ 7 \\ \hline & 2\ \ \ \\ \hline & {}_{8} \end{array}$$

first 4 x 7 = 28, so put down 2 carry 8.

The next step gives the product 9 x 7 = 63 and so we find 63 + 80 = 143.

Now here we have a 3-figure number, and that means the 1 in 143 must be carried to the 2 we have just written down, which then becomes a 3:

$$\begin{array}{rl} & 4\ 9 \\ \times & \ \ 7 \\ \hline & 2\ {}^{1}4\ 3 \\ \hline & {}_{8} \end{array} = \mathbf{343}.$$

Here we write the 1 in 143 as a superscript, and when it is added to the 2 we get the answer as 343.

With written calculations (it isn't a problem with mental calculations) you will sometimes need to increase a figure you have already written down.

This can be easily avoided however and that is explained in Chapter 9.

Let's take a similar example: **68 x 6**.

```
    6 8
x   6
  3
   6
```

First we get 36 as shown above.

Then we get 48, and adding the carry we get:
48 + 60 = 108.

So the 1 in 108 is added to the 3 already written down and we get:

```
    6 8
x   6
  3 ¹0 8 = 408.
   6
```

Next, multiplying a 3-figure number: **767 x 3**:

```
    7 6 7
x   3
  2 2
   1 8
```

The first two steps are shown above.
The last step is: 7 x 3 = 21, 21 + 80 = 101.
This means the left-hand 1 in 101 is written as a superscript and is carried to the left:

```
    7 6 7
x   3
  2 2 ¹0 1 = 2301.
   1 8
```

Similarly, for **642 x 8** we initially get 48:

$$\begin{array}{rl} & 6\;4\;2 \\ \times & \underline{\;\;8\;\;\;\;\;} \\ & \underline{4_{8}\;\;\;\;\;\;\;} \end{array}$$

Then we have 32 + 80 = 112.

So carry 1 to the left put down 1 and carry 2 forward, like this:

$$\begin{array}{rl} & 6\;4\;2 \\ \times & \underline{\;\;8\;\;\;\;\;} \\ & \underline{4_{8}\,{}^{1}1_{2}\;\;\;} \end{array}$$

Finally 16 + 20 = 36 which we put down:

$$\begin{array}{rl} & 6\;4\;2 \\ \times & \underline{\;\;8\;\;\;\;\;} \\ & \underline{4_{8}\,{}^{1}1_{2}\,3\,6} = \mathbf{5136}. \end{array}$$

The above multiplication is different from the others we just did because the 3-figure number is written with superscript/answer digit/**subscript** (that's because there's another product still to be done).

... numbers are the thoughts of God.....the Divine Wisdom is reflected in the numbers impressed on all things.....

St Augustine

Practice

a $\begin{array}{r} 64 \\ \times\ \underline{8} \end{array}$

b $\begin{array}{r} 78 \\ \times\ \underline{7} \end{array}$

c $\begin{array}{r} 38 \\ \times\ \underline{9} \end{array}$

d $\begin{array}{r} 484 \\ \times\ \underline{6} \end{array}$

e $\begin{array}{r} 545 \\ \times\ \underline{7} \end{array}$

f $\begin{array}{r} 455 \\ \times\ \underline{7} \end{array}$

Answers: **a** 512 **b** 546 **c** 342 **d** 2904 **e** 3815 **f** 3185

MULTIPLYING 2-FIGURE NUMBERS

Now let's apply this left to right written method to the product of two 2-figure numbers.

We'll use the vertical and crosswise pattern for this as we did in Chapter 5.

>>

Suppose we want **43 x 65**.

$$\begin{array}{rl} & 4\ 3 \\ \text{x} & \underline{\ 6\ 5} \\ & \underline{2_4} \end{array}$$

First we multiply vertically on the left: 4 x 6 = 24, as shown above.

Next we multiply crosswise and add: 4x5 + 3x6 = 38.

To this 38 we add the carried 4 (as 40): 38 + 40 = 78, so we put down 7 carry 8:

```
    4 3
x   6 5
  2 7
   4 8
```

Finally we multiply vertically on the right: 3 x 5 = 15. And 15 plus the carried 8 (as 80) gives 15 + 80 = 95, which we put down:

```
    4 3
x   6 5
  2 7 9 5
   4 8
```

So **43 x 65 = 2795**.

It's just vertical – crosswise – vertical, with the carries multiplied by ten.

Jedediah Buxton (1702–1772) from Derbyshire, England was once asked to find the amount obtained by doubling a farthing 140 times. His answer consisted of a 39-figure number for the pounds plus 2s 6d. Asked to multiply this 39-figure number by itself he gave the answer after two months having calculated it mentally from time to time over that period.

Another example: **91 x 82**.

```
    9 1
x   8 2
  7 4 6 2
   2 6
```

First we get 72, put down 7 carry 2.

Then 9x2 + 1x8 = 26, 26 + carried 2 (as20) = 46, put down 4 carry 6.

Finally 1 x 2 = 2, 2 + carried 6 (as 60) = 62, put down 62.

For **18 x 27**:

```
    1 8
x   2 7
  0 4 8 6
   2 3
```

First 1 x 2 = 2, put down 0 carry 2 (but you don't have to write down the zero).

Then crosswise gives 23, 23 + 20 = 43, put down 4 carry 3.

Finally 56 + 30 = 86, which we put down.

So **18 x 27 = 486**.

Practice

a	3 1	**b**	5 3	**c**	2 3	**d**	4 4
×	4 7	×	4 3	×	2 4	×	5 3

e	4 5	**f**	7 3	**g**	8 3	**h**	7 7
×	3 3	×	3 8	×	8 3	×	3 1

Answers: **a** 1457 **b** 2279 **c** 552 **d** 2332 **e** 1485 **f** 2774 **g** 6889 **h** 2387

3-FIGURE TOTAL

Now let's look at one where we get a 3-figure total and have to carry 1 leftwards: **57 x 75**.

$$\begin{array}{r} 5\ 7 \\ \times\ \ 7\ 5 \\ \hline 3\ \ \ \ \ \\ _{5}\ \ \ \ \ \end{array}$$

Initially we get 35 as shown above.
The crosswise step gives 25 + 49 = 74,
and 74 + 50 = 124.

The 1 in 124 must be taken to the left, so write it as a superscript::

$$\begin{array}{r} 5\ 7 \\ \times\ \ 7\ 5 \\ \hline 3\ {}^{1}2\ \ \\ _{5}\ \ _{4}\ \ \end{array}$$

The 2 is put in the answer and the 4 is carried forward.

Finally, vertically on the right we get 35, 35 + 40 = 75 which we put down.

$$\begin{array}{rl} & 5\ 7 \\ \times & \underline{\ 7\ 5\ } \\ & \underline{3\,^{1}2\ \ 7\ \ 5} = \mathbf{4275}. \\ & {\scriptstyle 5\quad 4} \end{array}$$

Let's see another example like that: **17 x 18**.

$$\begin{array}{rl} & 1\ 7 \\ \text{x} & \underline{\ 1\ 8\ } \\ & \underline{0\ 2\quad} \\ & {\scriptstyle 1\quad 5} \end{array}$$

First we get 1 and put down 0 carry 1.

Crosswise we get 15 and 15 + 10 = 25.

These two steps are shown above.

In the last step we get 56 and 56 + 50 = 106.
So the 1 in 106 is written as a superscript and the 06 is put down to complete the product:

$$\begin{array}{rl} & 1\ 7 \\ \text{x} & \underline{\ 1\ 8\ } \\ & \underline{0\ 2\,^{1}0\ 6} = \mathbf{306}. \\ & {\scriptstyle 1\quad 5} \end{array}$$

Practice

a 44 × 74	**b** 53 × 34	**c** 37 × 47	**d** 94 × 83

Answers: **a** 3256 **b** 1802 **c** 1739 **d** 7802

As already mentioned these 3-figure totals can be avoided by an ingenious and powerful but very simple device, which we explain next.

> *"For the harmony of the world is made manifest in Form and Number, and the heart and soul and all the poetry of Natural Philosophy are embodied in the concept of mathematical beauty."*
>
> ***D'Arcy Thompson***

8 BAR NUMBERS

As you are about to see, bar numbers are a very unusual but very powerful tool for mathematics. They allow any digits over 5 (i.e. 6, 7, 8 and 9) to be removed from a calculation if and when desired.

So they give us flexibility in the way we represent numbers.

Currently numbers can only be expressed in one way, but here you will see that any number can be expressed in different ways, and we can choose the one that's most convenient.

NEGATIVE TIMES

If we say the time is 8:55 we mean that the time is 55 minutes past 8 o'clock. But we may also say 'five minutes to nine' instead of 8:55.

'Five to nine' (or 'ten to twelve' etc.) seems a good way to describe times when the time is just **before** a whole number of hours.

There is however no way to write 'five minutes to nine' with numbers.

We write 'five minutes **past** nine' as 9:05.

Suppose we write $9{:}0\bar{5}$ for five minutes **before** nine.

BAR NUMBERS

The $\bar{5}$ is a 5 with a bar on top, signifying negative 5. In other words, instead of going 5 minutes **on** from 9 o'clock, $\bar{5}$ means we go 5 minutes **back**.
Similarly 12:10 means 10 minutes **past** 12, but 12:$\overline{10}$ means 10 minutes **to** 12, or 11:50.

It's simply a convenient way to describe these times.

NEGATIVE DIGITS

It is of course widely accepted that all the digits in a number are positive. So in the number 345 the digits 3, 4 and 5 are all positive.

But this does not have to be so. There's no reason why some of the digits in a number can't be positive and some negative.

For example the number **21** has two positive digits: the 2 represents two positive 10s and the 1 represents 1 positive unit.

> Now look at **2$\bar{1}$**.
>
> Here the 2 means 20 and the $\bar{1}$ means –1.
>
> So 2$\bar{1}$ means 20 – 1 or **19**.

2$\bar{1}$ is another way of representing the number 19.

Just as 9:0$\bar{5}$ means 5 minutes before 9 o'clock, 2$\bar{1}$ means 1 unit before 20, which is 19.

Similarly $\mathbf{3\bar{2}}$ means 30 – 2 or **28**.

And $\mathbf{10\bar{1}}$ means **99** (1 below 100).

This ingenious and effective device has been invented many times in the past and is a feature of Vedic Mathematics.

Numbers with negative digits have gone by various names, but we will just call them "bar numbers" because of the minus or bar that is placed on the top of digits that are to be counted as negative.

In fact we already use this in our speech. When we say, for example: "We are all here bar one" meaning everyone except one person is present, we are using exactly this idea.

Practice

Convert to ordinary numbers:

a $2\bar{2}$ **b** $7\bar{1}$ **c** $5\bar{3}$ **d** $6\bar{4}$

e $8\bar{3}$ **f** $10\bar{3}$ **g** $1\bar{2}$ **h** $3\bar{6}$

Answers: **a** 18 **b** 69 **c** 47 **d** 56 **e** 77 **f** 97 **g** 8 **h** 24

CREATING BAR NUMBERS

We can of course go the opposite way and put a number **into** bar number form.

Given **59** we see it's close to 60: it's 1 below 60.
So we can write it as $\mathbf{6\bar{1}}$.

(Actually 59 can also be written it as $1\overline{41}$ because 59 is 41 below 100.)

Similarly, **48** = $\mathbf{5\bar{2}}$ because 48 is 2 below 50.

Practice

Convert to bar numbers:

a 17 **b** 38 **c** 49 **d** 29

e 58 **f** 79 **g** 9

Answers: **a** $2\bar{3}$ **b** $4\bar{2}$ **c** $5\bar{1}$ **d** $3\bar{1}$ **e** $6\bar{2}$ **f** $8\bar{1}$ **g** $1\bar{1}$

So, any digit with a bar on top is a negative digit.

Let's look at numbers with more of a mixture of positive and negative digits.

For example, $\mathbf{3\bar{2}1}$ means 300 – 20 + 1 = **281**. The bar here is on the 2 so it counts as –20.

To change $3\bar{2}1$ to an ordinary number it's best to focus on the $3\bar{2}$ part first.

Since $3\bar{2}$ is 28, then $\mathbf{3\bar{2}1 = 281}$.

Similarly, $\mathbf{7\bar{3}4 = 674}$ (simply convert $7\bar{3}$ to 67).

And converting the other way....

181 can be written as $\mathbf{2\bar{2}1}$ because 18 = $2\bar{2}$.

Practice

Convert to ordinary numbers:

a $4\bar{2}3$ **b** $5\bar{1}1$ **c** $2\bar{3}4$ **d** $6\bar{4}2$

Convert to bar numbers:

e 191 **f** 382 **g** 575 **h** 261

Answers: **a** 383 **b** 491 **c** 174 **d** 562 **e** $2\bar{1}1$ **f** $4\bar{2}2$ **g** $6\bar{3}5$ **h** $3\bar{4}1$

9 MULTIPLICATION WITH BAR NUMBERS

Now let's see how we can use these bar numbers to make calculations easier.

In Chapter 7 we were multiplying **49 x 7** and started like this:

$$\begin{array}{r} 4\ 9 \\ \times \quad \underline{\ \ 7\ } \\ \underline{2\ \ \ \ }\\ {\scriptstyle 8}\ \ \ \end{array}$$

At the next step we get 9 x 7 = 63 and need to add this to 80.

So 63 + 80 = 143, we carry the 1 back to the 2 and put down 43 to get 343 as the answer.

The 3-figure number here, 143, is due to the 8 we carried forward (which then became 80).

But with our bar numbers (see previous chapter) we can now do this multiplication much more easily.

When we find 4 × 7 initially, we don't put 28 down because we notice that that the 8 will cause a problem (a 3-figure result).

We change 28 into $3\bar{2}$.

So we begin **49 x 7** by putting down 3 and carrying the $\bar{2}$ over like this:

$$\begin{array}{r} 49 \\ \times\ \ 7 \\ \hline 3 \\ \hline {\scriptstyle \bar{2}} \end{array}$$

Next we get 63 and add $\overline{20}$ (remember the carry becomes ten times bigger when working left to right) to it to get $63 + \overline{20} = 43$, which we put down:

$$\begin{array}{r} 49 \\ \times\ \ 7 \\ \hline 343 \\ \hline {\scriptstyle \bar{2}} \end{array}$$

We get **49 x 7 = 343** as before.

So we can avoid these 3-figure totals in this way.

"All the effects of nature are only the mathematical consequences of a small number of immutable laws."

Marquis de Pierre Simon Laplace

Now take **68 x 8**.

The first step gives 48 and the 8 is large: it will mean adding 80 at the next step.

A glance at the next step shows that it gives 8 x 8 = 64 and so 80 will be added to 64 giving a 3-figure total.

That a 3-figure total will come up is very quick to see in practice. And it tells us that we should put down $5\bar{2}$ in this case rather than 48 at the first step:

$$\begin{array}{r} 6\ 8 \\ \times\ \ \ 8 \\ \hline 5\ \ \ \ \\ \hline \end{array}$$
$\bar{2}$

Now in the next step we get 64 and we add $\overline{20}$ to this:
$64 + \overline{20} = 44$, which we put down:

$$\begin{array}{r} 6\ 8 \\ \times\ \ \ 8 \\ \hline 5\ 4\ 4 \\ \hline \end{array}$$
$\bar{2}$

So **68 x 8 = 544**.

Another example from Chapter 7 is **767 x 3**:

$$\begin{array}{rr} & 7\,6\,7 \\ \times & 3 \\ \hline & 2\,2 \\ \hline & {\scriptstyle 1\;\;8} \end{array}$$

The first two steps are done here. But notice the 8 carried: this can lead to a 3-figure result.

In fact the third step is 7 x 3 = 21 and 21 + 80 is a 3-figure number.

So we don't put down 2 and carry 8. Instead we put down 3 and carry $\bar{2}$ (because 28 = $3\bar{2}$).

We get:

$$\begin{array}{rr} & 7\,6\,7 \\ \times & 3 \\ \hline & 2\,3 \\ \hline & {\scriptstyle 1\;\;\bar{2}} \end{array}$$

The last step is 7 x 3 = 21, 21 + $\overline{20}$ = 1, so we put down 01. Remember we always put two figures down at the last step.

$$\begin{array}{rr} & 7\,6\,7 \\ \times & 3 \\ \hline & \mathbf{2\,3\,0\,1} \\ \hline & {\scriptstyle 1\;\;\bar{2}} \end{array}$$

As pointed out before, this can all be written out on a single line if you wish:

767 × 3 = $2_1 3_{\bar{2}} 01$ = **2301**.

Practice

a 38 × 9 **b** 87 × 6 **c** 46 × 7

d 654 × 8 **e** 484 × 6 **f** 478 × 4

Answers: **a** 342 **b** 522 **c** 322 **d** 5232 **e** 2904 **f** 1912

MULTIPLYING 2-FIGURE NUMBERS

When multiplying two 2-figure numbers we proceed in the same way: we anticipate those 3-figure totals and use the bar numbers if they come up.

So for **57 x 75**, covered in Chapter 7:

$$\begin{array}{lll} & 5 & 7 \\ \times & 7 & 5 \\ \hline & 3 & \\ & {}_{5} & \end{array}$$

Multiplying vertically on the left we mentally get 35. BUT, look at the crosswise result: we get 25 + 49 = 74. And with that carried 50 we will get a 3-figure total.

So, rather than 35 we put down $4\bar{5}$:

$$\begin{array}{lll} & 5 & 7 \\ \times & 7 & 5 \\ \hline & 4 & \\ & {}_{\bar{5}} & \end{array}$$

This is done mentally: you don't put the 35 down straight away, but check the next step to see if $4\bar{5}$ is better.

Then you know that the 74 we get at the next step will be reduced by 50:

$74 + \overline{50} = 24$ and we put down 2 and carry 4:

$$\begin{array}{lll} & 5 & 7 \\ \times & 7 & 5 \\ \hline & 4 & 2 \\ & {}_{\bar{5}} & {}_{4} \end{array}$$

In the last step: 7 x 5 = 35, 35 + 40 = 75, which we put down:

$$\begin{array}{lllll} & 5 & 7 & & \\ \times & 7 & 5 & & \\ \hline & 4 & 2 & 7 & 5 \\ & {}_{\bar{5}} & {}_{4} & & \end{array}$$

So **57 x 75 = 4275**.

MULTIPLICATION WITH BAR NUMBERS

Practice

a $\begin{array}{r} 44 \\ \times\ \underline{74} \end{array}$ **b** $\begin{array}{r} 53 \\ \times\ \underline{34} \end{array}$ **c** $\begin{array}{r} 37 \\ \times\ \underline{47} \end{array}$ **d** $\begin{array}{r} 96 \\ \times\ \underline{37} \end{array}$

Answers: **a** 3256 **b** 1802 **c** 1739 **d** 3552

> *Ernst Kummer was a well-known German algebraist. Though he was brilliant at algebra his arithmetic was often rather poor. Apparently, during one lecture he had to find 9×7: "Right, 9×7, that's er... um...". One student, seeing his difficulty, suggested it was 61, which Kummer wrote on the board. Kummer was about to continue when another student said he thought it should be 69. Then Kummer said "Come, come gentlemen, it can't be both. it must be one or the other."*

NO DIGITS OVER 5

As mentioned before, the digits 6, 7, 8 and 9 can be removed from a calculation.

And another way in which we may use the bar numbers is in changing one of the numbers being multiplied.

Suppose we want **33 x 39**.

Of course there is no problem using the method we have had. But that 9 is a large digit and we may prefer to work with smaller digits by using $4\bar{1}$ rather than 39.

Let's see what happens when we multiply **33 by** $\mathbf{4\bar{1}}$.

$$\begin{array}{lcc} & 3 & 3 \\ \times & 4 & \bar{1} \\ \hline 1 & & \\ \hline {}_{2} & & \end{array}$$

Vertically on the left 3 x 4 = 12, put 1 carry 2, as shown above.

Crosswise: $3 \times \bar{1} + 3 \times 4 = -3 + 12 = 9$.

9 + 20 = 29, put down 2 carry 9:

$$\begin{array}{lcc} & 3 & 3 \\ \times & 4 & \bar{1} \\ \hline 1 & 2 & \\ \hline {}_{2} & {}_{9} & \end{array}$$

Finally vertically on the right: $3 \times \bar{1} = \bar{3}$,
$\bar{3} + 90 = 87$, which we put down:

$$\begin{array}{lccc} & & 3 & 3 \\ \times & & 4 & \bar{1} \\ \hline 1 & 2 & 8 & 7 \\ \hline {}_{2} & {}_{9} & & \end{array}$$

So **33 x 39 = 1287**.

MULTIPLICATION WITH BAR NUMBERS

This brings out the versatility that the bar numbers give us. It takes a while to get used to them, but it pays off very well.

The bar numbers can be introduced anywhere in a calculation. So this really does give us a great deal of scope and flexibility.

Practice

a 49 × 33 **b** 43 × 38 **c** 59 × 54 **d** 59 × 34

Answers: **a** 1617 **b** 1634 **c** 3186 **d** 2006

10 SUBTRACTION

You may recall from Chapter 4 that we had two subtraction methods. Here we look at these again: the first is extended to a completely general method and the second is the same as the method given previously except that the carry figures are written down rather than held in the mind.

METHOD 1 – ALL FROM 9 AND THE LAST FROM

In Chapter 4 we saw how to do subtractions using the Vedic formula *All from 9 and the Last from 10*. In that chapter we mentioned that it leads to a completely general subtraction method, so we'll look at that now.

>>

Just a reminder first: suppose you need **4000 – 237**.

Remember that the first figure of 4000 is reduced by 1 to give the first figure of the answer as 3.

(This is because that 237 is coming off 1 of those 4 thousands so only 3 thousands will be left.)

The rest of the answer is obtained using the *All from 9 and the Last from 10* formula applied to 237. And that gives 763.

So **4000 – 237 = 3763**.

>>

GENERAL SUBTRACTION

Now suppose we want: **762 – 286**.

$$\begin{array}{r} 7\,6\,2 \\ -\ \underline{2\,8\,6} \\ \underline{} \end{array}$$

Subtracting in each column we get:
7–2 = 5,
6–8 = –2,
2–6 = –4.

Since the negative answers can be written with a bar on top we can write:

$$\begin{array}{r} 7\,6\,2 \\ -\ \underline{2\,8\,6} \\ \underline{5\,\bar{2}\,\bar{4}} \end{array}$$

And $5\,\bar{2}\,\bar{4}$ means 500 – 24 (see the example we just did immediately before this one) and
500 – 24 = **476**.

Children sometimes subtract numbers by subtracting in each column disregarding whether the top digit is greater than or less than the bottom digit.

So they can do that as long as they always take the bottom number from the top one.

Now let's take **5026 – 1234**.

Subtracting in each column we get:

$$\begin{array}{r} 5\,0\,2\,6 \\ -\,\underline{1\,2\,3\,4} \\ \underline{4\,\bar{2}\,\bar{1}\,\bar{2}} \end{array}$$

Now we just need to find 4000 – 212 using *All from 9 and the Last from 10.*

So **5026 – 1234 = 3788**.

Practice

a $\begin{array}{r} 5\,4\,3 \\ -\underline{1\,6\,8} \end{array}$ **b** $\begin{array}{r} 5\,6\,7 \\ -\underline{2\,7\,9} \end{array}$ **c** $\begin{array}{r} 8\,0\,4 \\ -\underline{3\,8\,8} \end{array}$

d $\begin{array}{r} 7\,3\,7 \\ -\underline{5\,5\,8} \end{array}$ **e** $\begin{array}{r} 6\,4\,1\,3 \\ -\underline{1\,8\,7\,8} \end{array}$ **f** $\begin{array}{r} 8\,0\,2\,4 \\ -\underline{5\,3\,3\,9} \end{array}$

Answers: **a** 375 **b** 288 **c** 416 **d** 179 **e** 4535 **f** 2685

The whole heaven is number and harmony"

Aristotle

A MIXTURE OF POSITIVE AND NEGATIVE

Now take **6767 – 1908**.

Subtracting in each column:

$$\begin{array}{r} 6\,7\,6\,7 \\ \underline{1\,9\,0\,8} \; - \\ 5\,\bar{2}\,6\,\bar{1} \end{array} \quad = \mathbf{4859}$$

We get $5\bar{2}6\bar{1}$ and the best way to remove the negative digits is to split the number where it changes from a negative digit to a positive digit:

$5\bar{2}/6\bar{1}$.

Now treat each part separately:
$5\bar{2} = 48$
and $6\bar{1} = 59$.

So $5\bar{2}/6\bar{1}$ = **4859**, which is the answer.

If a number has a mixture of positive and negative digits like this the best way is to **split the number wherever a negative digit is followed by a positive digit**.

Practice

a $\begin{array}{r} 6543 \\ -\ \underline{2818} \end{array}$ **b** $\begin{array}{r} 7143 \\ -\ \underline{3927} \end{array}$ **c** $\begin{array}{r} 4595 \\ -\ \underline{1748} \end{array}$ **d** $\begin{array}{r} 3274 \\ -\ \underline{2468} \end{array}$

Answers: **a** 3725 **b** 3216 **c** 2847 **d** 806

METHOD 2

Here you will see how to subtract numbers using the second method given in Chapter 4. In fact it's the same method but with the carry figures written down rather than held in the mind.

Suppose you want **63 – 37**.

You look in the left-hand column and subtract. You get 3. But before writing it down you look in the next column.

$$\begin{array}{r} 6\ 3 \\ -\ \underline{3\ 7} \\ \underline{} \end{array}$$

Seeing that you cannot take 7 from 3 you therefore put down 2 rather than 3, in the left column, and put the other one as shown.

$$\begin{array}{r} 6\ {}^{1}3 \\ -\ \underline{3\ \ 7} \\ \underline{2\ \ } \end{array}$$

Then the final step is just 13 – 7 = 6:

$$\begin{array}{r} 6\ {}^{1}3 \\ -\ \underline{3\ \ 7} \\ \underline{2\ \ 6} \end{array}$$

So **63 – 37 = 26**.

It doesn't matter if the carry figures are written as superscripts (as here) or subscripts. A superscript is used here to keep the carry away from the lower number, which makes it a little easier.

In this method you start at the left, subtract, and write this down if the subtraction in the next column **can** be done.

If it **cannot** be done you put down one less and carry 1 to the right, and then subtract in the second column.

Similarly, with **94 – 56** you look at the tens figures and subtract.

You get 9 – 5 = 4.

But since the subtraction in the units column cannot be done we give 3 as the first figure of the answer (1 less than 4), and append that 1 to the 4 in 94.

$$\begin{array}{r} 9\ {}^{1}4 \\ -\ \underline{5\ \ 6} \\ \underline{\mathbf{3\ \ 8}} \end{array}$$

Finally 14 – 6 = 8.

So **94 – 56 = 38**.

Practice

a $\begin{array}{r} 64 \\ -\ \underline{38} \end{array}$ **b** $\begin{array}{r} 44 \\ -\ \underline{29} \end{array}$ **c** $\begin{array}{r} 82 \\ -\ \underline{47} \end{array}$

d $\begin{array}{r} 91 \\ -\ \underline{38} \end{array}$ **e** $\begin{array}{r} 55 \\ -\ \underline{27} \end{array}$ **f** $\begin{array}{r} 73 \\ -\ \underline{57} \end{array}$

Answers: **a** 26 **b** 15 **c** 35 **d** 53 **e** 28 **f** 16

EXTENDING TO LONGER NUMBERS

To subtract longer numbers we simply extend this process. We always check the next column as we work through and put one less down if the next column cannot be done.

>>

Find **864 – 378**.

$$
\begin{array}{cccc}
 & 8 & {}^{1}6 & 4 \\
- & 3 & 7 & 8 \\
\hline
 & 4 & & \\
\hline
\end{array}
$$

In the left column we get 8 – 3 = 5 but we put down 4 because we cannot do 6 – 7 in the next column.

$$
\begin{array}{cccc}
 & 8 & {}^{1}6 & {}^{1}4 \\
- & 3 & 7 & 8 \\
\hline
 & 4 & 8 & \\
\hline
\end{array}
$$

Now for the middle column: 16 – 7 = 9.
But again put down one less, 8, because we cannot do 4 – 8 in the last column.

Finally on the right: 14 – 8 = 6 so the answer is **486**.

$$
\begin{array}{cccc}
 & 8 & {}^{1}6 & {}^{1}4 \\
- & 3 & 7 & 8 \\
\hline
 & \mathbf{4} & \mathbf{8} & \mathbf{6} \\
\hline
\end{array}
$$

>>

With a little practice it's really very easy and quick.

Practice

a	634 − 378	**b**	704 − 323	**c**	582 − 347
d	333 − 177	**e**	5665 − 2777	**f**	7003 − 5775

Answers: **a** 258 **b** 381 **c** 235 **d** 156 **e** 2888 **f** 1228

"Now, that's much better!"

CONCLUSION

This book demonstrates that calculating left to right for all the basic operations is possible. Whether it is better than right to left the reader can decide – hopefully after giving it a fair appraisal following careful practice.

But there is no doubt that left to right has certain definite advantages:

1) It's more natural and easier for mental calculation because we read and write from left to right.

2) It's more efficient because you can get just the figures you need instead of having to do the whole calculation every time. For example, to find the first two figures of 67,453 × 74,884 you would be looking for your calculator because it involves doing the whole calculation: two 5-figure numbers by long multiplication. But working left to right and using the *Vertically and Crosswise* method it is done quickly and easily in your head, as shown earlier in this book.

3) It's also more consistent because division is always done left to right and never right to left.

4) Sometimes you cannot start on the right. If you need to find a square root, solve a quadratic equation or find a trig (or inverse trig) value, there is no last figure to start at: you must start at the left.

The square root of 7, the solution to $x^2 + 5x - 7 = 0$, the value of sine 0.3 and so on (see Reference 1 for all these and more), all have no last figure for their answer, so you must start at the left.

You simply get the first (left-hand) figure, which is easy. Then you use that figure to get the next figure, and so on.

5) The other huge advantage of left to right calculation is that calculations can be combined. The marvellous possibilities that this opens up make it a very powerful technique.

Let's just see a couple of illustrations of combining operations.

COMBINING CALCULATIONS
Example 1

First suppose you want: **43 x 52 + 32 x 81**.

Look at it like this:

```
4 3        3 2
5 2   +    8 1
```

1) Starting at the left we obtain both vertical products and add them up:

```
4   3          3   2
|              |
5   2    +     8   1
```

(4x5) + (3x8) = 20 + 24 = 44, we write this as 4_4

So the left vertical step gives: 43 x 52 + 32 x 81 = 4_4

2) Next do the cross-products and add them:

```
4   3          3   2
  X              X
5   2    +     8   1
```

(4x2 + 3x5) + (3x1 + 2x8) = 23 + 19 = 42.

Now look at that 4_4 from the first step.

The subscript 4 means 40 and is to be added to the 42 we have just found.

This gives 42 + 40 = 82, which we write as 8_2

So we now have: 43 x 52 + 32 x 81 = 4_48_2

3) Finally we add the two right vertical products:

```
4   3          3   2
    |              |
5   2    +     8   1
```

(3x2) + (2x1) = 6 + 2 = 8.

Now add that carried 2 (as 20): 8 + 20 = 28, so 28 are the last two figures of the answer:

43 x 52 + 32 x **81 = $\mathbf{4_48_228}$ = 4828.**

This is how it works, and when necessary we can prevent numbers building up by using the bar digits.

COMBINING CALCULATIONS
Example 2

Finding the third side of a right-angled triangle given the other two sides involves squaring two numbers, adding them, and taking the square root of the result.

And it can be done from left to right, getting the answer, digit by digit until the required accuracy is obtained.

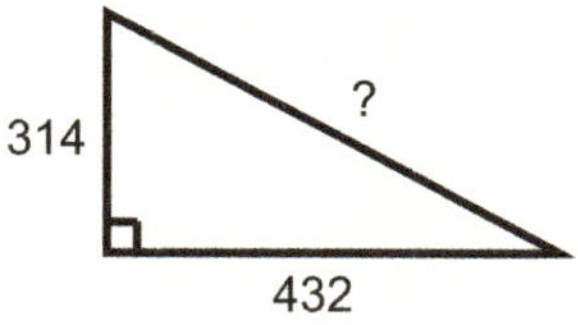

CONCLUSION

So the missing side above will be: $\sqrt{314^2 + 432^2}$.

And $\sqrt{314^2 + 432^2}$ = $5_0 3_0 4_1 . 0_6 6 \ldots$

where the carry figures are shown as subscripts. So the answer is **534.06** to 5 significant figures [1].

This shows the whole working, and it is not difficult once you know the one-line squaring and square rooting methods of Vedic Mathematics [6,7,8].

Combining calculations like this requires more memory power if carried out mentally, so written methods are useful. This way of combining calculations can be developed much further though and applies in the evaluation of trigonometric values (and their inverses), the solution of equations and so on [1,4].

This method should be taught in schools though, for the above reasons, instead of it being left to us, as adults, to work them out for ourselves, as many of us have done.

There is no suggestion here that right to left calculations should not be taught in schools: right to left may be the best way to introduce the basics of arithmetic.

REFERENCES

1. Williams K. R. Vedic Mathematics Teacher's Manual – Advanced Level. Inspiration Books, 1999. ISBN 978-1-902517-18-6.

2. Williams K. R. *Astronomical Applications of Vedic Mathematics.* Inspiration Books, 2000. ISBN 978-1-902517-22-3.

3. Williams K. R. *Triples.* Inspiration Books, 1999. ISBN 978-1-902517-19-3.

4. Nicholas A. P., K. Williams, J. Pickles. *Vertically and Crosswise.* Inspiration Books, 1984.
ISBN 978-1-902517-21-6.

5. Williams K. R. *The Natural Calculator.* Inspiration Books, 2000. ISBN 978-1-902517-15-5.

6. Sri Bharati Krsna Tirthaji, "Vedic Mathematics", published by Motilal Banarsidass, 1965. ISBN 81-208-0163-6.

7. Williams K. R. Vedic Mathematics Teacher's Manuals – Elementary and Intermediate Levels. Inspiration Books, 2002. ISBNs 978-1-902517-16-2 and 978-1-902517-17-9.

8. Williams K. R. *Discover Vedic Mathematics.* Inspiration Books, 1984. ISBN 978-1-902517-20-9.

APPENDIX 1

GROUPING

Grouping numbers in some comfortable way is a very useful technique, and one application is shown here. This follows on from Chapter 5.

If you need **123 x 132**, they are of course 3-figure numbers.

But if you think of the first two figures as a pair you can apply the same vertical/crosswise/vertical pattern to get the product.

We can split the numbers up into 12/3 and 13/2, treating the 12 and 13 as if they were single figures.:

12	3	vertically: $12 \times 13 = 156$,
13	2	crosswise: $12 \times 2 + 3 \times 13 = 63$,
162	3 6	vertically: $3 \times 2 = 6$.

Combining these mentally we get:

A. 156

B. $156_{6}3 = 1623$

C. 16236

So **123 x 132 = 16236**.

This grouping method works well when the first pair of figures of each number are easy to multiply.

Similarly with **203 x 103**.
Note that 20 x 10 is easy, so we group like this: 20/3 and 10/3.

A. 20 x 10 = 200.
B. 20x3 + 3x10 = 90. So 200$_9$0 = 2090.
C. 3 x 3 = 9. So 20909.

So **203 x 103 = 20909**.

EVEN BETTER

An even better method allows you to get the answer two figures at a time.

Suppose you want **304 x 402** as before.
And suppose you group the numbers differently.

Instead of 30/4 and 40/2 we can group the last two figures together like this: 3/04 x 4/02.

3	**04**
4	**02**

Note carefully the difference between this and what we did before, because when we group a pair on the right we get the answer two figures at a time, like this:

A. 3 x 4 = 12.
B. 3x2 + 4x4 = 22.
C. 4 x 2 = 08.

So we get **304 x 402 = 12/22/08 = 122208**.

Note that because each part has 2 digits we must put 08 for the last pair, not just 8.

This is particularly useful when the vertical product **on the right** can be done fairly easily.

Even 4-figure numbers can be multiplied like this.

If you want **1105 × 1103** you can split like this:
11/05 x 11/03.

Because there are two figures on the right we get the answer two figures at a time again.

11	05
11	03

A. 11×11 = 121.
B. 11×3 + 5×11 = 88. So we get 12188.
C. 5×3 = 15.

So we get 1218815.

And **1105 × 1103 = 1218815**.

One more example: **312 × 1011**:

3	12
10	11

A. 3x10 = 30.
B. 3x11 + 12x10 = 153. So $30_{1}53$ = 3153.
Note the carry here because 153 is a 3-figure number.
C. 12 x 11 = 132. So $3153_{1}32$ = 315432 (again we get a carry).

So **312 × 1011 = 315432**.

Given a choice about how to split the numbers, as in the last example, it's generally best to mark off two figures on the right and then work with pairs of figures. This way we tend to avoid the carry figures.

APPENDIX 2

THE MOVING MULTIPLIER METHOD

In multiplying a long number by a single figure, for example 4321 × 2, we multiply each of the figures in the long number by the single figure. We may think of the 2 moving along the row, from left to right, multiplying each figure vertically by 2 as it goes. The following method follows on from Chapter 5.

Suppose we want to multiply a 3-figure number by a 2-figure number: **531 x 24**.

First imagine the 24 in this position:

5	**3**	**1**
2	**4**	

Multiply vertically: 5 x 2 = 10. **Remember 10**.

Then multiply crosswise: 5x4 + 3x2 = 26.

Combine 26 with the 10 from before: $10_{2}6$ = 126.
Remember 126.

Next imagine the 24 in this position:

5	**3**	**1**
	2	**4**

Now multiply crosswise: 3x4 + 1x2 = 14.
Combine this with the 126 from before:
$126_{1}4$ = 1274.
Remember 1274.

And finally multiply vertically on the right: 1 x 4 = 4.
Combine this with the 1274 from before: 12744.

So **531 x 24 = 12744**.

If we want **4321 × 32**.
That is, a 4-figure number multiplied by a 2-figure number, the 32 takes three positions.

4 3 2 1 **3 2**	We put 32 first of all at the extreme left. Then vertically on the left, 4 × 3 = 12. And crosswise, 4×2 + 3×3 = 17. Combining these: $12_{1}7$ = **137**.
4 3 2 1 ** 3 2**	Then move the 32 along and multiply crosswise: 3×2 + 2×3 = 12. Combining with137 we get: $137_{1}2$ = **1382**.
4 3 2 1 ** 3 2**	Moving the 32 once again: multiply crosswise, 2×2 + 1×3 = 7. Combining this: **13827**. Finally the vertical product on the right is 1×2 = 2.

And combining with 13827 gives **138272**.

Therefore **4321 × 32 = 138272**.

So we multiply vertically on the left,
then crosswise in every position,
then vertically at the very end.

Finally take a 5-figure number: **31013 × 21**.

Here the 21 takes these four positions:

3 1 0 1 3 **2 1**	**3 1 0 1 3** **2 1** (under 1 0)	**3 1 0 1 3** **2 1** (under 0 1)	**3 1 0 1 3** **2 1** (under 1 3)

We multiply vertically on the left, then crosswise in the four positions, and then vertically on the right.

The six mental steps give: 6,5,1,2,7,3.

So the answer is **651273**.

For written calculations, as in Part 2 of this book we can proceed as follows.

"I do not know what I may appear to the world; but to myself I seem to have been only like a boy playing on the seashore, and diverting myself in now and then finding a smoother pebble or a prettier shell than ordinary, whilst the great ocean of truth lay all undiscovered before me."

Sir Isaac Newton

Suppose we need: **731 x 24** and we want to write out the calculation.

```
    7 3 1
x   2 4
    1
     4
```

As before we line the multiplier (24) to the left so that the 2 is under the 7.

Then multiplying vertically on the left we get 7 x 2 = 14, put down 1 carry 4.

Now multiply crosswise: 7x4 + 3x2 = 34.
And 34 + the carried 4 (as 40) = 74, put down 7 carry 4:

```
    7 3 1
x     2 4
  1 7
   4 4
```

Now move the 24 one place to the right and cross-multiply: 3x4 + 1x2 = 14.
And 14 + the carried 4 (as 40) = 54, so put down 5 carry 4:

```
    7 3 1
x     2 4
  1 7 5
   4 4 4
```

Finally multiply vertically on the right: 1 x 4 = 4,
4 + carried 4 (as 40) = 44, which we put down.

```
    7 3 1
x     2 4
  1 7 5 4 4
   4 4 4
```

You can make it a little easier by re-writing the multiplier (24 in the last example) every time you move it. And you can write it above or below the number (wherever there is space).

In this example, we get a 3-figure total and need to carry 1 to the left: **654 x 32**.

$$\begin{array}{cccccc} & & 6 & 5 & 4 & \\ \times & & 3 & 2 & & \\ \hline & 1 & & & & \\ \hline & & {\scriptstyle 8} & & & \end{array}$$

Initially we get 18.
The next step gives 27 and 27 + 80 = 107.

So the 1 in 107 is written as a superscript:

$$\begin{array}{cccccc} & & 6 & 5 & 4 & \\ \times & & 3 & 2 & & \\ \hline & 1^{1} & 0 & & & \\ \hline & & {\scriptstyle 8} & {\scriptstyle 7} & & \end{array}$$

The rest of the calculation is straightforward:

$$\begin{array}{cccccc} & & 6 & 5 & 4 & \\ \times & & & 3 & 2 & \\ \hline & 1^{1} & 0 & 9 & 2 & 8 \\ \hline & & {\scriptstyle 8} & {\scriptstyle 7} & {\scriptstyle 2} & \end{array} = \mathbf{20928}.$$

So **654 x 32 = 20928**.

Alternatively we can use the bar numbers, writing $2\bar{2}$ instead of 18 at the first step:

$$\begin{array}{r} 6\,5\,4 \\ \times \quad 3\,2 \\ \hline 2\,0\,9\,2\,8 \\ \hline {\scriptstyle \bar{2}\ \ 7\ 2\ \ } \end{array}$$

Question: What do the numbers 3, 24, 34, 41 and 59 have in common?

Answer: They all come with fried rice!

APPENDIX 3

EXTENDING THE PATTERN

Here we show how the *Vertically and Crosswise* pattern for multiplying 2-figure numbers can be extended to the multiplication of 3 and 4-figure numbers. Once you understand the pattern you will see how it extends further.

Suppose you want to find **504 × 321**.

```
    5   0   4
    3   2   1
  -----------
  1 6 1 7 8 4
```

For 3-figure numbers there are five steps.

The pattern now becomes:

```
 5   0   4
 |
 3   2   1
----------
15
```

A Vertically on the left, 5×3 = 15.

B Then crosswise on the left,
5×2 + 0×3 = 10.
Combining the 15 and 10:
$15_{1}0 = 160$.

```
 5   0   4
   X
 3   2   1
----------
 1 6 0
```

C Next we take 3 products and add them up,
5×1 + 0×2 + 4×3 = 17. And $160_{1}7 = 1617$.
(actually we are gathering up the hundreds by multiplying hundreds by units, tens by tens and units by hundreds)

```
 5   0   4
     ✳
 3   2   1
----------
 1 6 1 7
```

D Next we multiply crosswise
on the right,
0×1 + 4×2 = 8: 1617,8 = 16178.

```
 5   0   4
       X
 3   2   1
----------
 1 6 1 7 8
```

E   Finally, vertically on the right,
4×1 = 4: 16178,4 = 161784.

```
  5  0  4
        |
  3  2  1
----------
1 6 1 7 8 4
```

Note the symmetry in the 5 steps:
first there is 1 product, then 2, then 3, then 2, then 1.

We may summarise these steps as follows:

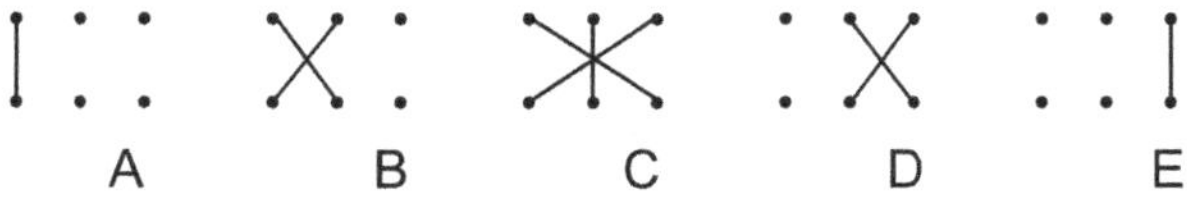

4-FIGURE NUMBERS

Now look at two 4-figure numbers: **3201 × 4302**.

```
   3  2  0  1
   4  3  0  2 ×
---------------
1 3 7 7 0 7 0 2
```

The seven steps are illustrated as follows:

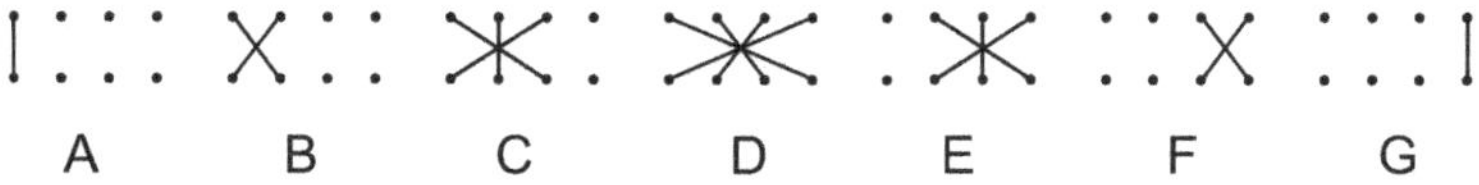

We get: A. 3×4 = 12
B. 3×3 + 2×4 = 17
C. 3×0 + 2×3 + 0×4 = 6
D. 3×2 + 2×0 + 0×3 + 1×4 = 10
E. 2×2 + 0×0 + 1×3 = 7
F. 0×2 + 1×0 = 0
G. 1×2 = 2

The mental steps are therefore:

$12_{1}7 = 137$
137,6 = 1376
$1376_{1}0 = 13770$
13770,7,0,2 = **13770702**.

Please see References 1, 4, 5, 6, 7, 8 for more on this.

INDEX

Made in the USA
Middletown, DE
05 November 2026

36544055R10068